HOW TO DELETE BOOKS IN KINDLE

The Ultimate Guide For Complete Beginners On How To Delete Books In Kindle In 5 Minutes.

I0481917

BY

ANDREW NIXON

Copyright©2018

COPYRIGHT

Andrew Nixon

TABLE OF CONTENT

THE END

CHAPTER 1

INTRODUCTION

The Kindle is one of the world most famous android tablet with over 3 million users who owns and uses it, with the amazing features it has in it.

Apart from the popular use of kindle device as an E-reader, it can also perform other functions as deleting books from your Kindle. The user interface is a friendly one and it has a lot of good stuffs in it.

This guide will show you how you can delete book from your Kindle; all you need to do is just follow this guide step by step as they are instructed. There are unlimited apps from other sources that can be deleted from your kindle tablet, but if you find this very difficult, there is no needs to worry because I got you cover.

All over the world millions of people haven't been able to use and perform wonders with its features but dis book gives the breakdown of all solution to any problem you might encounter.
Thankfully each steps are very easy and simple to follow, that even a beginner can master it in a few minutes.

CHAPTER 2

HOW TO DELETE BOOKS FROM KINDLE

An important merit of eBook reading device is the capacity, you can hold large number of eBooks with this handy size device with a basic version of Kindle. Although every coin has two sides, after you must have finished and achieved countless number of books, the storage space becomes likely disordered. Meanwhile, when you need some time to find the book you want to read, then it's time you get your ebook shelf cleaned. But before doing so, you should ensure to know the type of deletion you want.

1. You downloaded a lots of books and have completed most of them, now you need to make your Kindle clear and tidy, **delete those completed books from device but they should be kept in cloud.**

→ making use of Kindle device, like Kindle Paperwhite, Kindle Fire, Kindle Voyage, Kindle Touch

→ making use of Kindle app like, Kindle for Android and Kindle for IOS

2. You even don't want to see those already completed books in archive, wish to **delete them from Kindle Cloud.**

→ completely delete books from your Kindle cloud, Kindle archive.

DELETING BOOKS FROM KINDLE

The model with 4 buttons at the bottom and a 5-way controller. Presently delete books from Kindle is the easiest and cheapest kind in the Kindle family and it costs only $69.

To get your Kindle space free up, you can do that by selecting the item's name on the home screen and tapping on the left arrow on the 5-way controller. Then remove from device should be selected, and you should click on the center of the 5-way controller to get it done.

CHAPTER 3

STEPS TO TAKE TO GET BOOKS FROM KINDLE KEYBOARD DELETE

1. Press the home button, if you are not already on the home screen.

2. The list of the content that are already on your device, you should move the 5-way to underline the item that you wish to remove.

3. Drag the 5-way to the left to remove the item you underlined.

4. Books only that are purchased from the Kindle store you will see, "Remove from Device" but for all other content you are going see "Delete" press on the 5-way to

remove the content. If you wish to change your mind, drag the 5-way up or down to cancel. It should be noted that even if you books from your device, Kindle books are backed up for you at the Amazon.

5. You will need to confirm the deletion of the content by choosing "OK", for content other than books that you have purchased from the Kindle store.

CHAPTER 4

GET BOOKS FROM KINDLE PAPERWHITE DELETE

To get your library cleaned up on Kindle paperwhite, you can do this by pressing and holding the item's name or cover on the home screen. At the time when the dialog box pops up, select "Remove" from device. Note that this option varies depending on the content, and your content will remain stored safely in the cloud for a later time download.

CHAPTER 5

DELETING BOOKS FROM KINDLE VOYAGE

Deleting books from Kindle voyage is not complicated, all you need to do is just press and hold the title that you wish to remove for a little time, when a menu appear, choose "Remove from Device", then on your Kindle voyage the book will be deleted. You can also download this book from Kindle cloud if you wish to read it again.

CHAPTER 6

DELETING BOOKS FROM KINDLE OASIS

Maybe you wish to delete a recent book from Kindle oasis, move to your home screen, or to your library or search box to look for the item(s) that you wish to delete, just tap and hold it for a while until a box pops up with different options, choose "Delete this Book".

By doing so the content transported to the Kindle oasis via USD will be permanently deleted. But deleted books purchased from Kindle store or pushed to Kindle via email can be resync.

CHAPTER 7

DELETING BOOKS FROM KINDLE FIRE (HD)

To enable you delete books from Kindle Fire, all you need to do is just press and hold on an item to show the contextual menu, then choose remove from device. If for any reason you wish to download content purchased from Amazon for a later time, it remain stored in the Amazon Cloud.

CHAPTER 8

HOW YOU CAN GET BOOKS ON KINDLE DEVICES IN BATCH DELETE

Recently Amazon brought out a new firmware update 5.9.2.0.1 for its Kindle eReaders. Kindle Voyage (KV), Kindle (Kindle 7), Kindle (Kindle 8), Kindle Paperwhite 3 (KPW 3), Kindle Paperwhite 2(KPW 2), Kindle Oasis 2 (KO 2), Kindle Oasis (KO), all support the new firmware. This new firmware enable you get rid of Kindle books in batch.

To get rid of Kindle books in batch, move to; Home,
Menu,
Settings,
Device Options,
Advanced Options,
Storage Management,
Manual Removal.

After that your device storage can be manage by selecting the items you would like to delete.
Or this can also be achieved by just moving to; Home,
Menu,
Settings,
Device Options,
Advanced Options,
Storage Management,
Quick Archive, to quickly remove item from this device that has not been unveil for past 1/3/6, or 1 year.
Amazon which adds storage management to the new firmware as a small function, it's really comfortable for Kindle users that want their storage frequently clean.
Compare this to manual removal, quick archive will make your beloved content easy by accident. But please do not try this method if you are unable to remember all the contents in the Kindle local disk.

If you wish to delete books from Kindle devices in batch, you can upgrade the firmware to the latest 5.9.2.0.1 version. And software firmware can be updated to 5.9.2.0.1 version on your Kindle device or download firmware 5.9.2.0.1 directly from Amazon.
NOTE: All steps above just got rid of the books from your device and they are archive to your cloud. If you wish to permanently delete them, you have to remove them from the Kindle cloud.

CHAPTER 9

DELETING BOOKS FROM KINDLE APP FOR ANDROID AND IOS

Get Item From Kindle Android App Remove

You have to enter the application first before removing content from your Kindle app that are installed on your Android device. Once in the home screen your books can be display there.

Locate the book that you wish to remove from your cell phone in the app, then you should tap and hold on the book you wish to get rid of, then select "Remove From Device".

Get Item From Kindle IOS App Remove

In order not to use your device's memory space to get books stored, books can be delete from your iPhone, iPod, or iPad device.

To enable you remove individual book from the Kindle for iOS app, all you need to do is just tap and hold the cover of the book, then select "Remove From Device". Note that for all iOS 7 version Kindle app, an extra "Add to Collection" option will be there. Book synced from cloud, the appear choice will be "Remove from Device". And for book that only exist on your device, its choice will be "Delete Permanently".

CHAPTER 10

DELETING BOOKS FROM YOUR KINDLE CLOUD

Ensure to understand that before deleting a book permanently from your Kindle cloud, you will be unable to have access or read it again unless the book is been purchase again.

note that by deleting books from your Kindle cloud means you don't own it anymore.

Now lets delete item from Kindle cloud permanently with step by step as shown below:

1. **First You Should Log In To Amazon.com, Then Move to "Manage Your Content and Devices"** (formerly known as "Manage Your Kindle")

Here, all content in your Kindle cloud can be seen and can be sync to all your registered apps and devices.

2. Choose the book that you wish to delete, then tap on "Delete" button.

3. Then a warning sign will appear to enable you make sure. Tap "Yes".

Now, finally this book have been deleted permanently and you will unable to view or see the book in your device or cloud again.

THE END

Andrew Nixon

Andrew Nixon

Andrew Nixon

www.ingramcontent.com/pod-product-compliance
Lightning Source LLC
Chambersburg PA
CBHW072034230526
45468CB00021B/1787